Ocean Life
An Adult Coloring Book

Copyright © 2021 Robber Fickle
All Rights Reserved.

Copyright © 2021 Robber Fickle
All Rights Reserved.

All rights reserved. No part of this publication may be reproduced, distributed, or transmitted in any form or by any means, including photocopying, recording, or other electronic or mechanical methods, without the prior written permission of the publisher, except in the case of brief quotations embodied in critical reviews and certain other noncommercial uses permitted by copyright law.

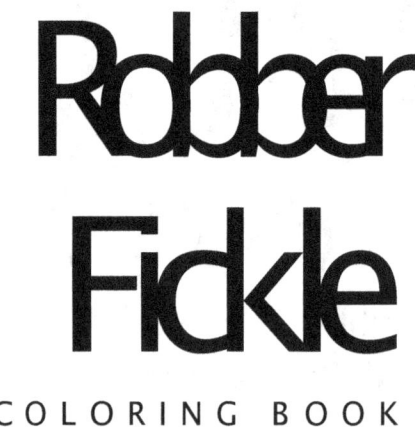

COLORING BOOK

For Any Question and Suggestions
robberfickle@gmail.com

@robberfickle

www.ingramcontent.com/pod-product-compliance
Lightning Source LLC
Chambersburg PA
CBHW080442220526
45465CB00007B/2733